We dedicate this book to the travelers of all sizes, to biodiversity, and to these incredible creatures, the humpback whales—may they forever journey freely in the ocean.

JOURNEY OF THE HUMPBACKS

Written by
Juliana Muñoz Toro

Illustrated by
Dipacho

translated by Lawrence Schimel

EERDMANS BOOKS FOR YOUNG READERS
GRAND RAPIDS, MICHIGAN

This is an invitation for us to go on a journey.
We don't need a backpack or shoes. Just our eyes wide open.
We'll meet the ocean's largest acrobats, and we'll travel with them from the glacial waters near Antarctica to the warm coasts of the Pacific, back to the place where everything began, where they were born.

Here begins the fascinating migration of *Megaptera novaeangliae*: the journey of the humpback whales.

Beyond the southernmost tip of land on the planet, beneath a gray sky that seems to hang down low over the earth, there is a channel of ocean that lies between Antarctica and a chain of islands. This is the Gerlache Strait. Here there are peaks of rocks covered in snow, reflected in a calm sea where small islands of ice float. Below this watery mirror is another world, one where even the heaviest animal can fly.

In the Antarctic silence, it's hard to believe that a whole world of movement exists underwater. Penguins, sea lions, and orcas dance their routines silently in the icy water.

But then—a sound. A deep breathing at the surface of the water. Two blowholes appear: the sign of a gigantic animal whose ancestors once walked on land. It dives, showing first its humped back and then its tail, covered with unique patterns and scars like a fingerprint. This is a humpback whale.

Scientists call these creatures *Megaptera novaeangliae*, meaning "giant-winged New Englanders," referring to the whale's long fins and early sightings of these animals off the coast of New England. Those long fins—the whale's "wings"—are necessary to fly through the water the way humpbacks do. Whales also leap into the air (or breach) to orient themselves, to court, to compete, to communicate, or to play. A few belly pirouettes can even help them clean off the barnacles which attach to their skin. Humpbacks are the most acrobatic of all whales, always jumping and playing.

Gerlache Strait, Antarctica

Southern Ocean

Coordinates: 64°30'00"S 62°20'00"O

Distance traveled: 0 kilometers (0 miles)

The humpbacks begin their journey in March or April. This species feeds in one hemisphere and crosses the equator to reproduce and bear their young in the waters of another hemisphere. Some travel in groups, but they may separate along the way. Others follow their route as solitary giants.

HUMPBACK WHALE *(Megaptera novaeangliae)*

DORSAL FIN

This single fin on the whale's back stabilizes its swimming, so it doesn't turn sideways. The dorsal fin and tail are like fingerprints for a whale: each animal has a uniquely shaped dorsal fin, and unique patterns on its tail. Researchers use these characteristics to identify each individual.

SKIN

Whale skin feels like hard rubber or a tire.

TAIL

The tail moves up and down, helping it push its body forward and move through the water.

HAVE YOU WONDERED?

1. How large can a humpback grow?
   They can grow up to 18 meters (59 feet) long.
2. Are male or female whales larger?
   Female humpbacks are up to 1.5 meters (4.9 feet) bigger than males.
3. What do they smell like?
   They smell like shrimp and fish.
4. How long do they live?
   They live for nearly 60 years.
5. Is there any other whale with fins larger than the humpback?
   No. Humpback whales have the longest pectoral fins of all whales. That's why the first part of their scientific name means "giant-winged."

PECTORAL FINS

These fins, located on each side of the whale, can measure up to 6 meters (19.7 feet)—a third of the length of its body.

EMPEROR PENGUIN

The emperor penguin is the largest species of penguin that exists. These birds can grow up to 1.2 meters (4 feet) long, and they weigh between 20 and 40 kilograms (44 to 88 pounds). They only live in Antarctica.

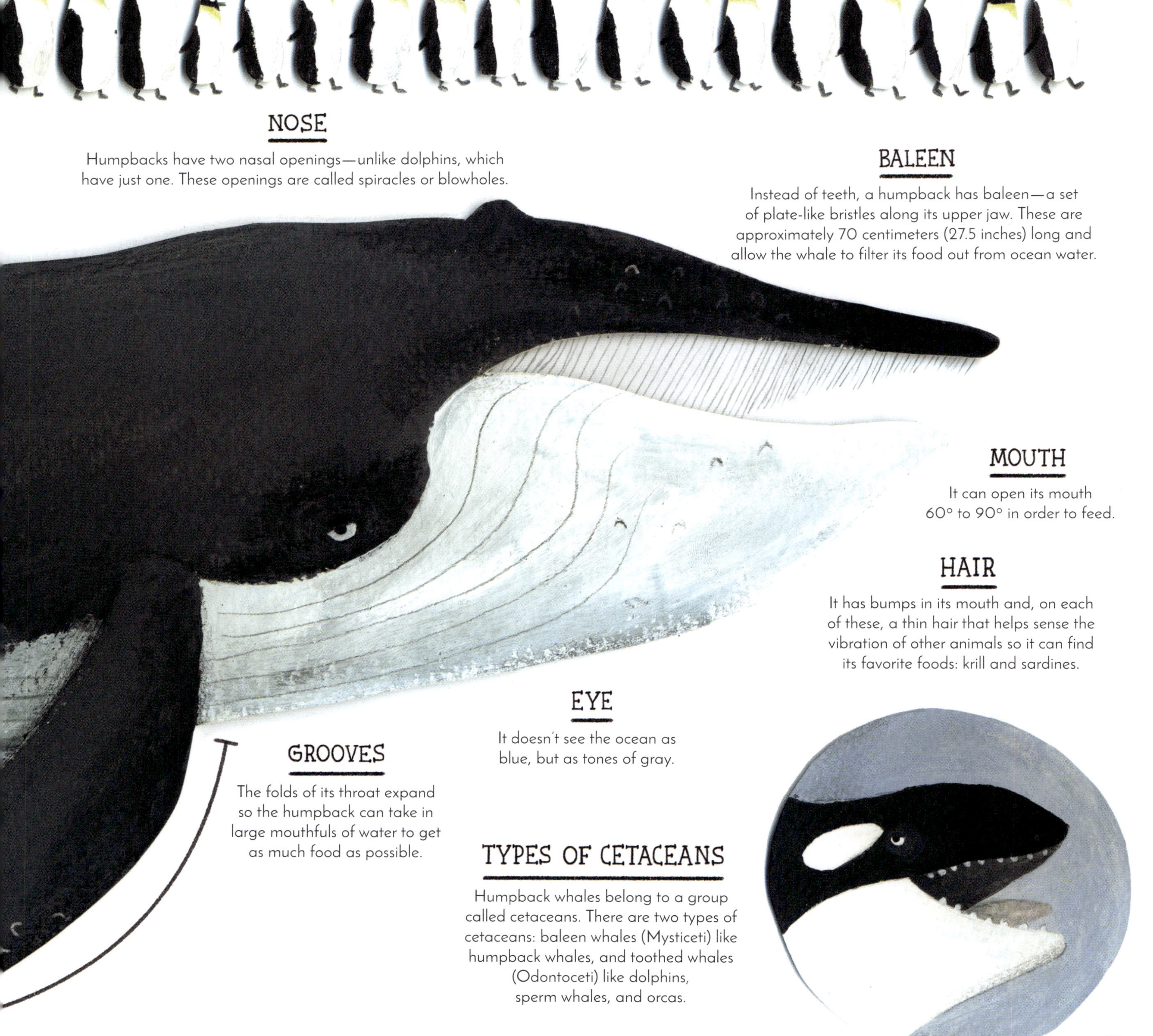

NOSE

Humpbacks have two nasal openings—unlike dolphins, which have just one. These openings are called spiracles or blowholes.

BALEEN

Instead of teeth, a humpback has baleen—a set of plate-like bristles along its upper jaw. These are approximately 70 centimeters (27.5 inches) long and allow the whale to filter its food out from ocean water.

MOUTH

It can open its mouth 60° to 90° in order to feed.

HAIR

It has bumps in its mouth and, on each of these, a thin hair that helps sense the vibration of other animals so it can find its favorite foods: krill and sardines.

EYE

It doesn't see the ocean as blue, but as tones of gray.

GROOVES

The folds of its throat expand so the humpback can take in large mouthfuls of water to get as much food as possible.

TYPES OF CETACEANS

Humpback whales belong to a group called cetaceans. There are two types of cetaceans: baleen whales (Mysticeti) like humpback whales, and toothed whales (Odontoceti) like dolphins, sperm whales, and orcas.

When they're hungry, the whales gather together. They swim toward the depths in circles while exhaling air, creating powerful "bubble nets" which rise toward the surface and catch small fish and crustaceans. The whales take turns feeding on these shoals with their mouths wide open. Water passes through their baleen, and only the food is left.

These large-finned swimmers are part of the natural balance. Nothing that happens in the waters does so in isolation. Whales are the best fertilizers of the ocean, and their abundant waste actually helps fight climate change! The nutrients released in the humpbacks' feces help many other organisms thrive—including many types of plankton. These plankton absorb carbon dioxide, one of the major causes of climate change. So whales both give and receive, taking the food they need but also keeping the ocean healthy for many other creatures.

These whales will eat one last meal before beginning the long journey to the place where they were born—where their own offspring and great-offspring will be born. As if they were packing a suitcase, the humpbacks have been filling their bodies with as much fat as possible. They won't eat again until they return to the Southern Ocean.

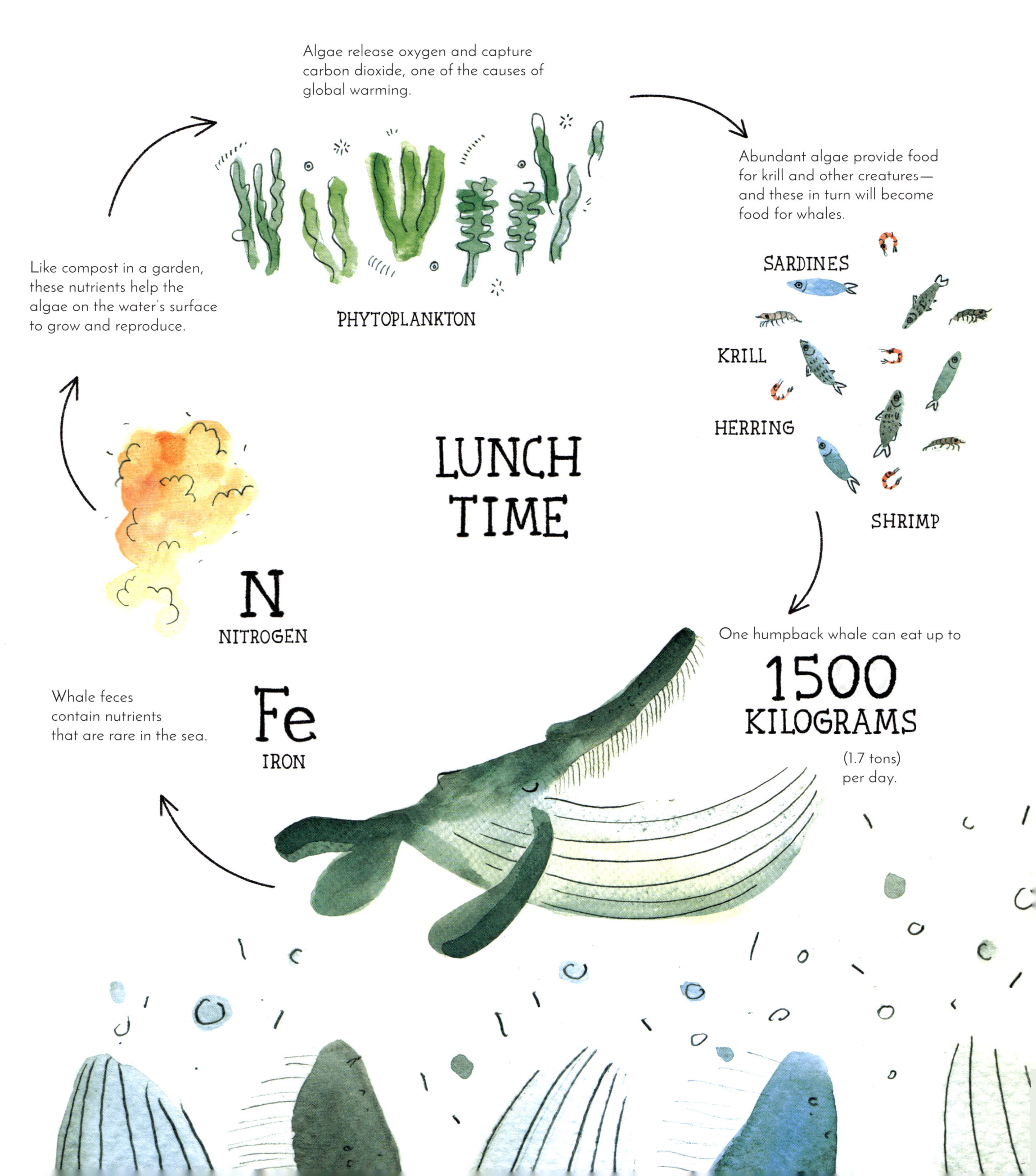
LUNCH TIME
Algae release oxygen and capture carbon dioxide, one of the causes of global warming.
PHYTOPLANKTON
Abundant algae provide food for krill and other creatures—and these in turn will become food for whales.
SARDINES
KRILL
HERRING
SHRIMP
One humpback whale can eat up to
1500 KILOGRAMS
(1.7 tons) per day.
Like compost in a garden, these nutrients help the algae on the water's surface to grow and reproduce.
N
NITROGEN
Fe
IRON
Whale feces contain nutrients that are rare in the sea.

The temperature is dropping lower and lower, and there are fewer hours of light in the Gerlache Strait, but the changing seasons are not the only reason the humpbacks decide to leave. It is also tradition. These whales were born in the tropics, and since then they have made the journey back each year. And now the mothers will show the next generation how to travel between their two homes. The journey begins, and their only guide will be memory.

TIME FOR ADVENTURES

The migrations of humpback whales are the longest of any mammal (other than humans), starting in March as autumn arrives in the Southern Hemisphere, lasting three to five months. Their migration is a path full of adventures. There are good days, like when a pod of dolphins liven up the voyage with their leaps and games. The juvenile whales, always trusting, chase after their cetacean cousins for a while. The adult humpbacks, with their grand size and peaceful spirit, have no enemies in the deeps, but orcas and sharks might attack a humpback calf. Danger is part of a young humpback's life. Some juveniles are hit by large ships, while others—for lack of experience—will wind up tangled in fishing nets or beached on the shore during very low tides.

NAP TIME

Do humpbacks get tired of swimming so far and so long to get back to where they were born? Will they look for a bed of warm currents and a blanket of algae?

When they're tired, humpbacks don't sleep deeply. Unlike humans, whales think about each breath they take—so if they were fully asleep, they might drown. They rest by shutting down half their brain for a few minutes, floating like ships in calm waters.

TIME TO BREATHE

The breathing of humpback whales is long and deliberate, as if they were meditating. From a short distance away, one can see the clouds that spring from their heads. These clouds are not seawater, but hot air that comes from their lungs, condensing into tiny water droplets when it comes in contact with the cold air. This is like going outside on a crisp day and seeing your breath become visible.

After whales take a breath, they close their nasal passages tightly so that no water gets in when they dive. They empty their lungs, moving the oxygen into their blood and muscles, which allows them to descend to the depths of the marine world. After 5 or 10 minutes—or sometimes even as long as 45 minutes—they move their tail like a motor and break the surface to breathe once more. They might remain topside for a while, to show off their acrobatic skills.

The humpback whales pass alongside the coastline of Peru. The water feels less cold here. Some females are accompanied by males looking for mates, or by other female whales who help protect last year's calves from predators. Other humpbacks migrate alone, each at its own pace. They are animals who come together for short periods during reproduction or feeding seasons. They will only eat when they return to the Southern Ocean near Antarctica, and they will only find a partner when they reach the tropics.

Although they are often far away from one another, humpbacks communicate among themselves with clicks: an oceanic voice that's intermittent, like the light from a lighthouse blinking in the dark. Sound travels faster underwater than it does in the air, and without the noise of ships and underwater construction, they can hear one another from ocean to ocean. Male humpbacks also sing—without vocal cords, and without needing to exhale. The air vibrates inside their bodies, like music inside a church. They repeat tones sometimes for hours, and different types of songs can follow trends and patterns among other humpbacks. The songs are a cultural revolution: the males from each group of whales sing the same song, made up of groans, wails, and squeaks. The song that Colombian humpbacks sing, for example, is different from that of Australian humpbacks. Males sing to catch the attention of a female, to scare off other males, or to define their territory—or perhaps just because they can.

The warm tropical waters already feel close. It's time to push toward the last leg of the journey. These long-finned whales slip gracefully through their world of underwater choruses. They are protectors, especially mothers with their offspring. Researchers have even seen some whales keeping company with others who are tangled in nets. Hopefully the humpbacks will keep singing, and humans will one day be able to understand what they are saying.

Peruvian Coast, South America

Pacific Ocean

Coordinates: 3°59'00"S 80°59'00"O

Distance traveled: approximately 6,500 kilometers (4,000 miles)

Mothers with their calves travel about 5 kilometers (3.1 miles) per hour. The more competitive males can reach up to 25 kilometers (15.5 miles) per hour. They almost never make stops—they want to reach their destination.

When a humpback whale is perhaps 60 years old, it is ready to say goodbye. But that's not the end of the story. Some would say that the whale dies; others, that new life is about to begin. When a whale passes away, its body descends to the seafloor—an environment with few nutrients—and it becomes a home and a meal for a large community of microorganisms.

Sharks, hagfishes, crabs, and other crustaceans are the first to arrive looking for food. When only the whale's large skeleton remains, then come bacteria and brittle stars (relatives of the starfish) to devour what they can from the decaying bones. Nothing is wasted. There are species of worms and snails that can easily digest bones. If the whale was beached on the shore, its body would feed birds and land animals. It is a feast.

If only all whales passed away from old age
and not because of the dangers of the journey.
Fortunately, hunting humpback whales has been
outlawed in most parts of our planet since 1986,
removing some of the danger to these creatures.
Yet still, here among the waves, the whale's body
is a reminder that nothing is permanent.

Land ahoy! How do the humpbacks know that they've reached their home? Perhaps because the waters are warm and the coasts are shallow. Perhaps because of the constant drops of rain splattering across the ocean's surface. (On the lush, green Pacific coast of Colombia, it rains almost every day of the year.) Our best answer, though, is that humpback whales have excellent memories.

Here in the tropics, some whales will be looking for mates. Other females have come with calves already in their bellies. Female humpbacks give birth after eleven and a half months of pregnancy, once they've reached the gentle currents of the Pacific Ocean near Colombia. Born tail-first, humpback calves weigh around 1,500 kilograms (about 1.7 tons) and measure up to 4 meters (13 feet) long. As these newborns arrive, it is August, and the world above the waves smells of damp forest.

A newborn whale calf is clumsy and uncoordinated: its body tilts, it floats like a balloon, and it can't dive easily. Its mother gently corrects it, using her body to keep her child in place. After a few days, the calf has become a bit more agile and begins to leap. The calf approaches boats with a curiosity that worries the mother. She uses a fin to discourage it.

Everything is a lesson for the calf, and these lessons can be the difference between surviving or not on the long journey back to the Southern Ocean that will start in about three months. The calf will need to gain weight, drinking its mother's milk, which is thick like sour cream. The mother, who still hasn't eaten since she left the Southern Ocean, will wait until the last moment to set out on the return trip, so that her calf has as much time as possible to grow. They won't separate until a year later, after she has shown her calf the route to the Gerlache Strait and then back to the Pacific where it was born. This migratory route is the most important school of a humpback whale's life.

Utría National Natural Park,
Colombia, South America

Pacific Ocean

Coordinates: 5°59'00"N 77°21'00"O

Distance traveled: 8,300 kilometers (5,150 miles)

Between July and November, the humpbacks arrive at Bahía Málaga, Gorgona, and Utría National Natural Parks—the humpback nursery. Humpback whales also travel to Ecuador, Panama, and Costa Rica to breed.

REPRODUCTIVE LIFE

1 Humpback whales may have several different partners during their lifetime.

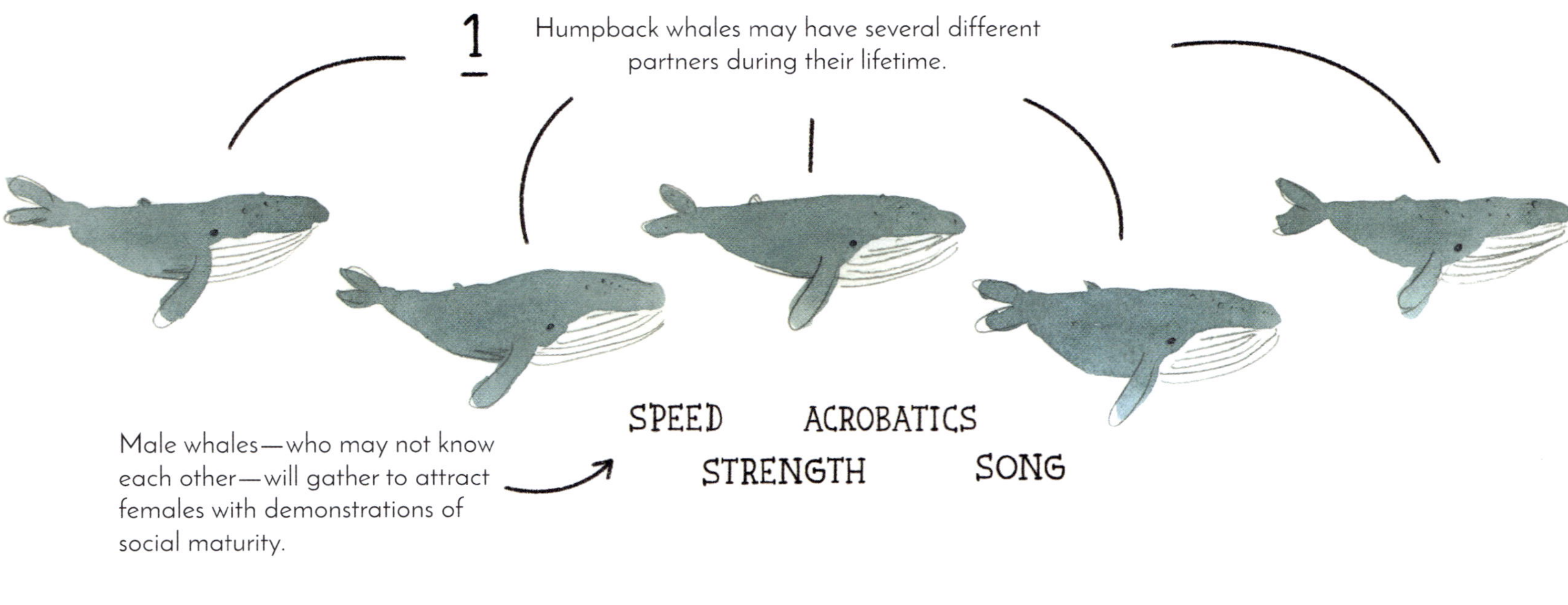

Male whales—who may not know each other—will gather to attract females with demonstrations of social maturity.

SPEED ACROBATICS STRENGTH SONG

2 Whale songs are acts of courtship—that is, the males sing to attract females and to mark their territories.

The males of each group of whales all sing the song of the season. It's like the "Hit of the Summer." The following year, the song changes.

ONLY MALE HUMPBACKS SING.

3 All whales leap, and the males do this to demonstrate their dexterity. There is a maneuver called "inversion" in which a male launches himself toward the sky and then falls back on his side. This is intended to make him seem larger than his rivals.

4 As they compete to mate with a female, the males challenge one another—sometimes even doubling their swimming speed.

THEY CAN RACE FROM 12 TO 25 KILOMETERS (7.5 TO 15 MILES) PER HOUR!

5 After several days of courtship, the female will join a male to mate. She may or may not mate with more than one male.

6 If the female has a calf at her side, she will try to reject the males courting her in order to protect her calf.

7 Very little is known about the mating behavior of humpback whales. It happens very quickly, so it has been difficult for scientists to study.

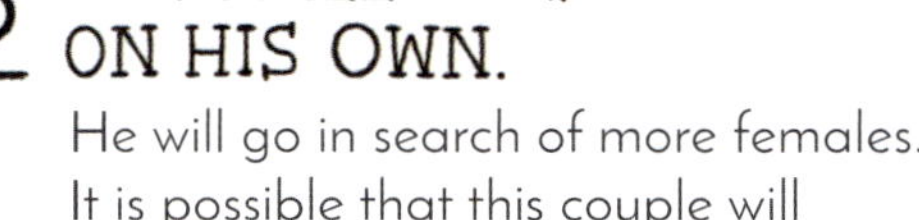

8 THE MALE GOES OFF ON HIS OWN.

He will go in search of more females. It is possible that this couple will never meet again.

When she migrates, the female will always return to her birthplace.

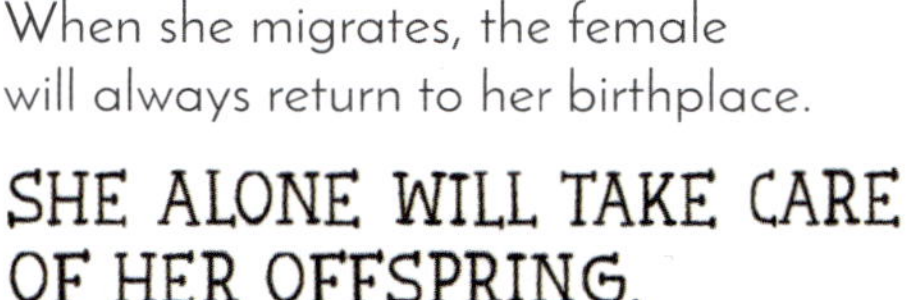

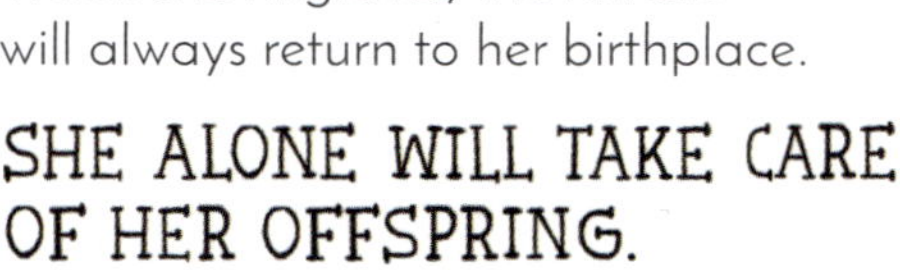

SHE ALONE WILL TAKE CARE OF HER OFFSPRING.

HAVE YOU WONDERED?

1. How long is a female humpback pregnant before giving birth?
   11.5 months (350 days).

2. How many calves can a female humpback have in her lifetime?
   Around 15. Normally she has one calf every two or three years.

3. How does a whale calf identify its mother?
   By her sounds. Scientists have recorded the sounds the calves use to communicate with their mothers.

On this shore along the Pacific Ocean, the houses rise from the ground on stilts due to constant flooding. Here children play on the sand, and at lunchtime the air smells like stews, fried bananas, and seafood. A marimba sounds. A downpour begins. At night, the frogs croak and luminous plankton shine in the water like floating constellations. In the early hours of the morning, motorboats arrive to collect travelers who have come here to see the whales.

In less than twenty minutes, the boat's driver spies the clouds of vapor these giants produce when they surface to breathe. It's possible to approach a little closer, but it's important to be careful and avoid bothering them with engine noise. With a little patience and luck, one might see a whale jumping out of the water, showing its long fins, with all its size and splendor. Who could see the whales like this, wild and free, without feeling a sense of joy and respect?

The humpbacks will make this long journey each year until the end of their days. The males might explore other destinations, but the females will faithfully follow their route from the Southern Ocean to the Pacific Ocean and back. It is their tradition, and they will pass it from generation to generation as long as these waters remain peaceful and clean. The humpbacks will always work to find their home—a home we share with them. They are our mirrors in the sea.

THE BALEEN WHALE FAMILY

Blue Whale
(Balaenoptera musculus)
30 meters (98 feet) long

It is the largest animal on land or in the sea.

Greenland Whale
(Balaena mysticetus)
18 meters (59 feet) long

It does not have a dorsal fin like other species in its family.

North Atlantic Right Whale
(Eubalaena glacialis)
18 meters (59 feet) long

Thousands of whales were killed and nearly driven to extinction during the whaling era that lasted from the 17th century until 1986, when commercial whaling was banned by the International Whaling Commission. The right whale was one of the most hunted whales because it is very strong and has a great amount of meat.

Humpback Whale
(Megaptera novaeangliae)
18 meters (59 feet) long

It is the protagonist of this book.

Gray Whale
(Eschrichtius robustus)
15 meters (49 feet) long.

It is relatively slow and likes to live near coastal areas.

Fin Whale

(Balaenoptera physalus)
21 meters (69 feet) long

Like the humpback, it also lives in the Pacific Ocean near Colombia.

Northern Sei Whale

(Balaenoptera borealis)
16 meters (52.5 feet) long

It is an internationally protected species because it almost became extinct when whaling was still allowed.

Southern Right Whale

(Eubalaena australis)
16 meters (52.5 feet) long

Like the humpback, it also migrates to warm waters during the southern winter. It has calluses on its head and the tip of its snout where some crustaceans live.

Bryde's Whale

(Balaenoptera brydei)
14 meters (46 feet) long

Another whale that lives in the Pacific Ocean.

Minke Whale

(Balaenoptera acutorostrata)
8 meters (26 feet) long

It is the most abundant species of whale in the world.

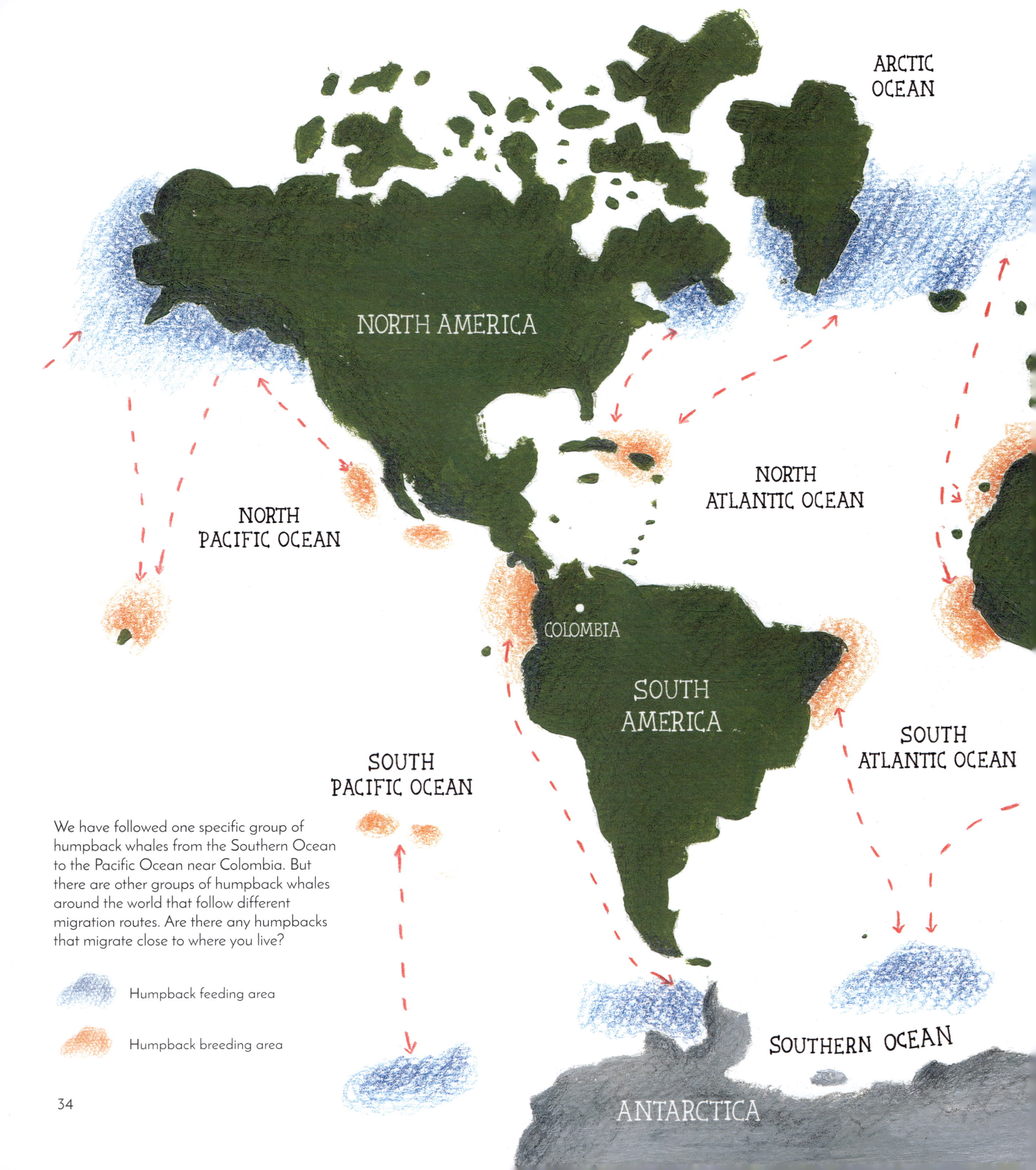

We have followed one specific group of humpback whales from the Southern Ocean to the Pacific Ocean near Colombia. But there are other groups of humpback whales around the world that follow different migration routes. Are there any humpbacks that migrate close to where you live?

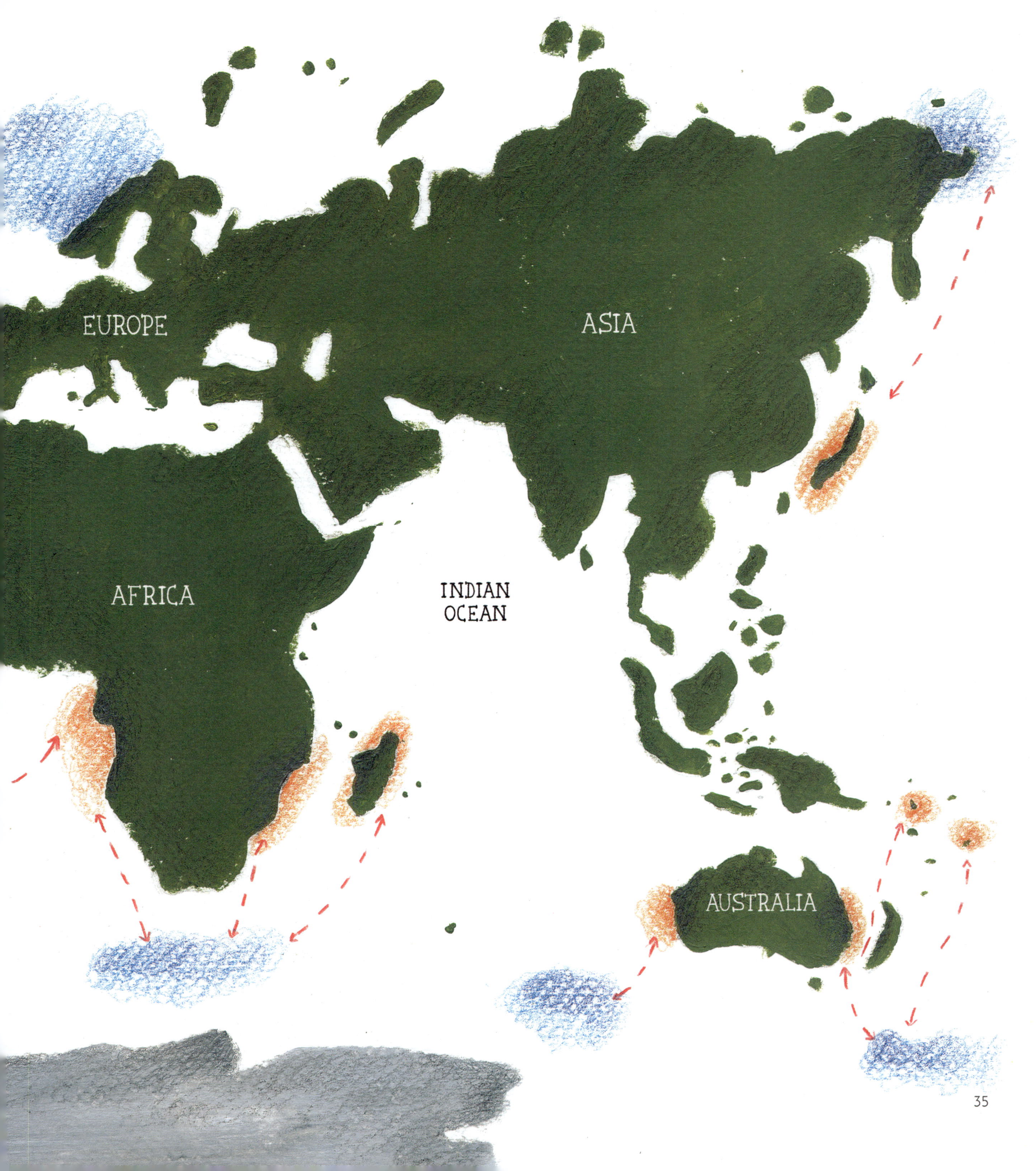
EUROPE
ASIA
AFRICA
INDIAN
OCEAN
AUSTRALIA

THIS BOOK

WAS MADE BY A TEAM OF PEOPLE
who love whales

CREATORS

DIPACHO

Children's Book Author & Illustrator

As a child, Dipacho was addicted to geography. He knew all the capitals and flags, and traveled the world via his atlas. When he grew up, he began to visit those countries he had dreamed about. He says that traveling stimulates his creativity. Then he stays in one place for a while to turn his life experiences into the books he makes, such as *Antonia* and *Some Do, Some Don't* (both Astra). Dipacho also designs board games and plays various musical instruments. He is self-taught. If he were a humpback, he would learn to sing . . . at last! Dipacho lives in Colombia. Visit his website at dipacho.com or follow him on Instagram @dipacho.

JULIANA MUÑOZ TORO

Writer

Juliana lives in Bogotá, Colombia, in a blue house full of carnivorous plants and miniature gardens. When she was little, she wrote everywhere: on the walls, under the chairs, or in little notebooks she made for herself. When she grew up, she became the author of over a dozen books, including *Journey of the Humpbacks*, her English-language debut. When she is not in front of the keyboard, Juliana is standing on her head or walking in the mountains. Because she is afraid of drowning, she probably wouldn't want to be a whale, but she loves to visit them in the Pacific. Visit Juliana's blog at julianamunoztoro.wordpress.com and follow her on Instagram @julianadelaurel.

RESEARCHERS

LILIÁN FLÓREZ-GONZÁLEZ

Marine biologist and director of the Fundación Yubarta (fundacionyubarta.blogspot.com), a Colombian organization that promotes aquatic biodiversity through research and environmental education

Lilián has been studying the aquatic mammals of Colombia in the wild for over 30 years, especially humpback whales. She is fascinated by their long migrations, their leaps, and their various ways of communicating. Although Lilián lives in the city of Cali, Colombia, she spends the majority of her time in the Pacific Ocean studying whales and other marine mammals. When she travels, she leads activities with local communities to raise awareness about environmental protection.

SUSANA CABALLERO GAITÁN

Biologist, microbiologist, and professor

She lives in the mountains of Bogotá with a large family of different species: two children, two cats, three dogs, and a hedgehog. She has spent almost her entire life looking for whales and dolphins. If she were a humpback whale, she would cross the equator to meet the humpbacks of the north. When she is not getting seasick in the boat or becoming a lab rat, Susana draws and dances flamenco and Arab dances. When she was little, she spent her time saving animals.

NATALIA BOTERO ACOSTA

Biologist and director of the Fundación Macuáticos Colombia (fundacionmacuaticoscolombia.wixsite.com), a multidisciplinary organization dedicated to growing human knowledge and conservation of aquatic life, especially aquatic mammals

As a girl, Natalia was obsessed with dinosaurs. But since dinosaurs are extinct, she devoted herself to studying the behavior and the areas where the humpbacks range. When she goes to photograph them in the North Pacific, Natalia admires the unique color patterns of their tails, while she enjoys the contrast between the jungle and the sea of Chocó. She finds it fascinating that enormous whales eat such tiny prey as krill.

TRANSLATOR

LAWRENCE SCHIMEL

Lawrence has written or translated over 300 books, including *Niños*, *9 Kilometers*, and the Batchelder Honor book *Different* (all Eerdmans). His works have received many awards, including three PEN Translates Awards, two SCBWI Crystal Kite Awards, and two Américas Award Honors. Lawrence lives in Madrid, Spain. Follow him on Bluesky @lawrenceschimel.bsky.social.

Originally published in Colombia as *El vuelo de las jorobadas*

English-language translation published by arrangement
with Base Tres Agency (base-tres.com)

First published in the United States in 2025 by Eerdmans Books for Young Readers,
an imprint of Wm. B. Eerdmans Publishing Co. • Grand Rapids, Michigan
www.eerdmans.com/youngreaders

 • Manufactured in China

34 33 32 31 30 29 28 27 26 25 1 2 3 4 5 6 7 8 9 10

ISBN 978-0-8028-5643-2

MIX
Paper | Supporting responsible forestry
FSC www.fsc.org FSC® C144853

A catalog record of this book is available from the Library of Congress.

Illustrations created with mixed media

Eerdmans Books for Young Readers would like to thank Jill Holz (B.S. Geology and Geophysics, M.Ed., and National Geographic Certified Educator) for sharing her scientific expertise for the English-language edition of this book.

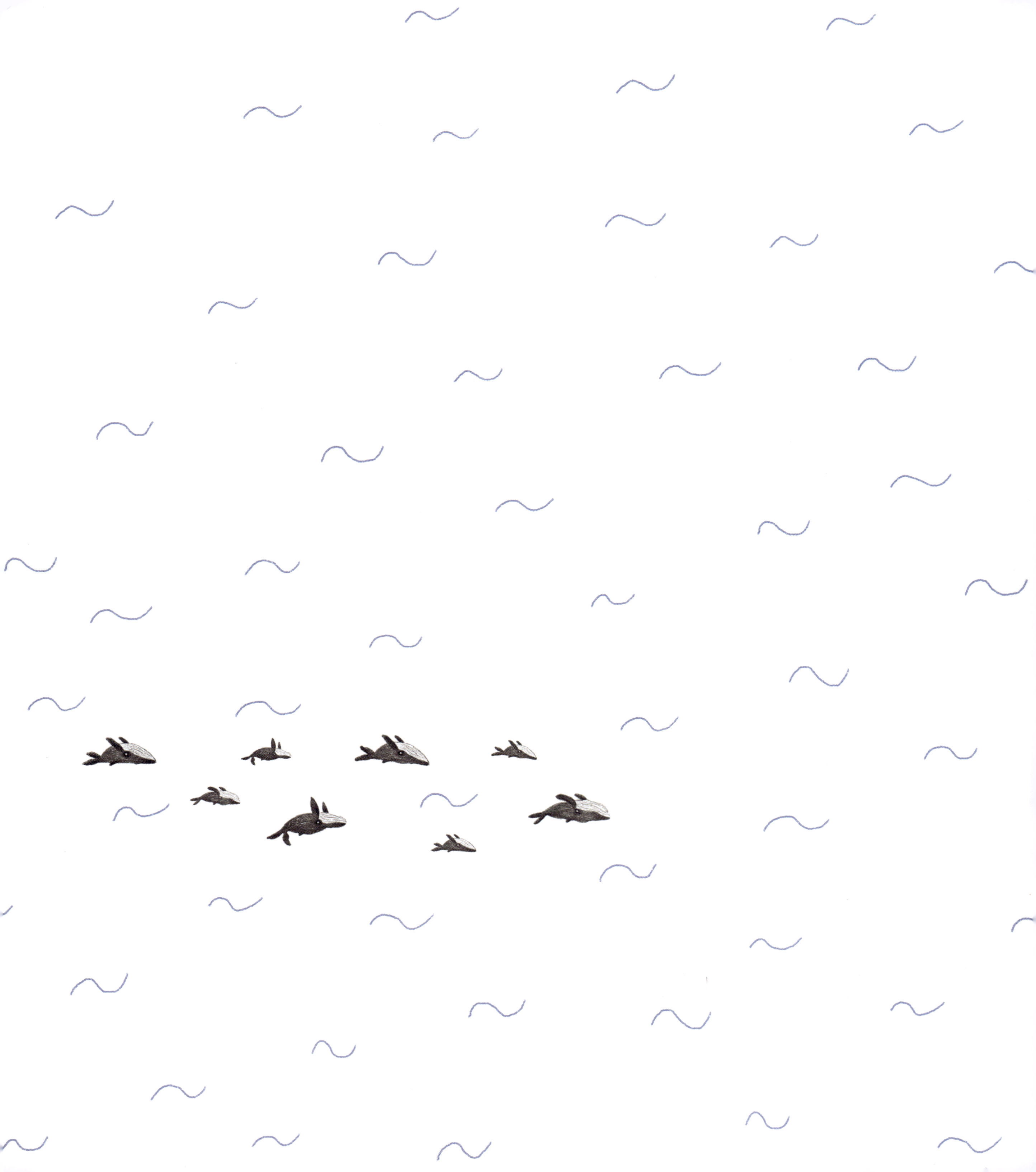